Math Concept Reader

# A New Angle on Trains and Train Stations

BY SARAH MASTRIANNI



Printed in China

ISBN 13: 978-0-15-360305-1
ISBN 10: 0-15-360305-4

9 10 11 0940 16 15 14 13
4500409990

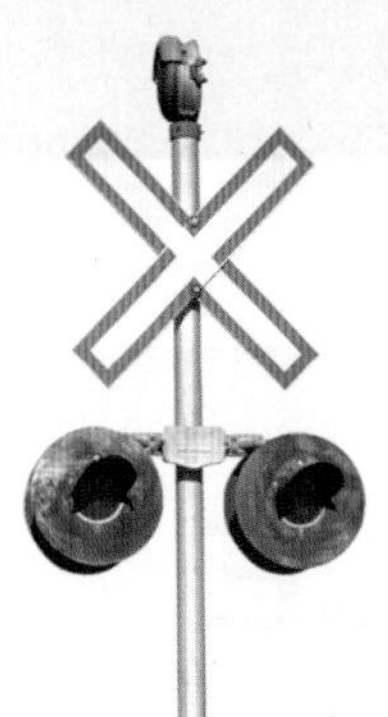

## CHAPTER 1:

# ANGLES, ANGLES EVERYWHERE

Before cars, buses, and airplanes, people traveled long distances by train. Trains run along rails on tracks. They are powered by steam, diesel fuel, or electricity. They have carried passengers and freight from one place to another for more than 100 years.

Many people may not be familiar with the train station. Train stations are also called railway stations. They are large, small, and nearly every other size. No matter the size, railway stations are usually busy with many people and lots of activity.

**A train station is one place to see geometric figures.**

Train stations provide a stop for both passengers and freight. Each station has information about trains and the times they stop. This information is called a schedule. Passengers can find the schedule on paper and on the Internet. Schedules are also on monitors and large boards at the stations.

Train tracks run north, south, east, west, and every direction in between. Trains run between towns, cities, and states, and across the country. Trains carry more than freight and passengers, though. They also carry geometric figures! A geometric figure is any set of points on a plane or in space. Lines, angles, polygons, and curves are examples of geometric figures.

**This is Union Station in Kansas City, Missouri.**

Math class is one place to study geometry; a train station is another. You can see designs that look like lines and angles in a station. You can find plane figures there, too. A plane figure is a figure that lies in one plane. A triangle is an example of a plane figure.

Almost everything you see at a train station contains geometric figures. Many stations are built with designs that look like figures. Take a look at the photo on this page. The roof and windows look like geometric figures. Can you see figures in other places in the image?

An angle is one type of geometric figure. The large clock at Grand Central Terminal in New York City forms new angles every minute. At 3:00, the clock hands make an L. At 1:00, the hands form a lesser angle; at 5:00 the hands form a greater angle.

The angle formed at 3:00 on the clock is called a right angle. The angle at 1:00 is called an acute angle. The angle formed at 5:00 is called an obtuse angle.

**The hands of the clock at Grand Central Terminal in New York City form different angles depending on the time shown.**

**This is the inside of Grand Central Terminal in New York City. There are many angles to see inside the building.**

There are many places to see right, acute, and obtuse angles at Grand Central Terminal. Look at the rectangular window frames in this photo. They form right angles at each corner. Even the smaller glass panes form right angles.

Look at the counters and schedule boards around the ticket windows. They also form angles. Look closer and you may see angles on the information booth. Do you notice the angles made by people and their shadows?

Trains out on the tracks have angles, too. In this photo, the train windows form angles. Even the painted design on the engine and cars forms angles. Look at the train and tracks. Can you find any right angles? Can you find any acute or obtuse angles?

What other geometric figures can you find on a train or in a station? One answer is lines! In math, lines go on forever in both directions. There are designs that look like lines in the train photo on this page. These designs are straight like lines, but they have an endpoint. Can you see some designs that look like lines in the photo?

**Look for acute, obtuse, and right angles on the train and the tracks.**

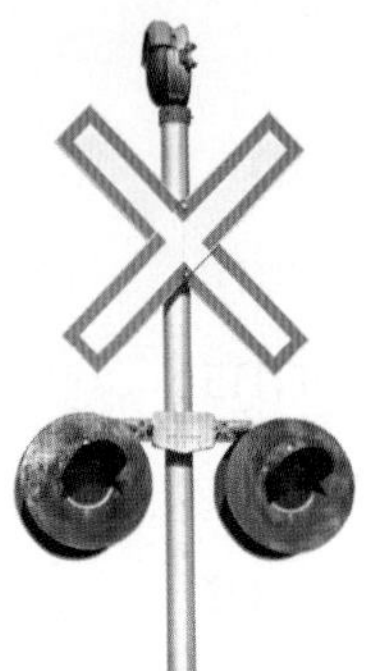

## CHAPTER 2:

# LINES — THEY JUST KEEP GOING

Some lines are called parallel. Parallel lines stay exactly the same distance apart. Some train tracks run parallel to one another. When we talk about tracks in this book we mean two metal rails and the wooden ties underneath the rails. Engines pull cars along the track by riding the rails. Trains can run on parallel tracks and never cross paths.

The two metal rails that run along the inside of a track never cross each other. The rails are always parallel. They are the same distance apart even if the track curves. The width between the rails is called the gauge. The United States standard railroad gauge is 4 feet, 8.5 inches.

Two lines that cross are called intersecting lines. The point where these lines meet is the point of intersection. Train maps show many points of intersection. They also show designs that look like parallel and intersecting lines.

Here, the Pink and Green Routes are parallel. The trains on these routes never intersect. The Pink Route and the Red Route do intersect. Their tracks meet at North Way, the point of intersection. Wilson Boulevard is the point of intersection for the Green, Blue, and Red Routes.

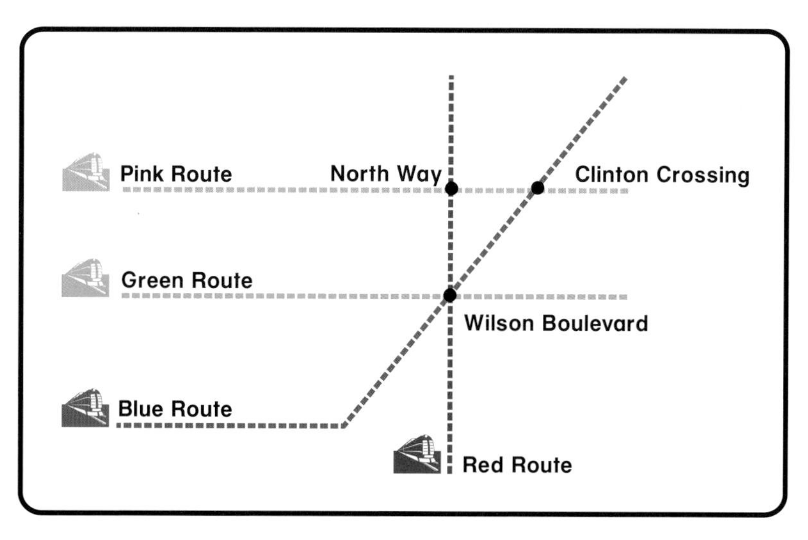


**Parallel, perpendicular, and intersecting lines are shown on this train map.**

Some intersecting lines are called perpendicular lines. Where the lines cross, they form four right angles. Railroad Crossing signs show perpendicular marks. You find these signs where a train track crosses a road. Tracks running north and south that intersect with tracks running east and west are perpendicular.

Look at the photo on this page. Can you find perpendicular designs on the train and the bridge? There are many perpendicular designs that remind us of lines on the buildings. Can you see them?

**Perpendicular lines can be found on the building and on the train bridge.**

**This is Union Station in St. Louis, Missouri.**

Lines and angles are everywhere in and around trains. "All aboard," to find even more geometric figures. Plane figures are nearly as plentiful as lines!

The photo on this page is of Union Station in St. Louis, Missouri. What plane figures do you see in the image? Look at the railing along the bridge. Now look at the steel beams at the top of the station. Try to find plane figures.

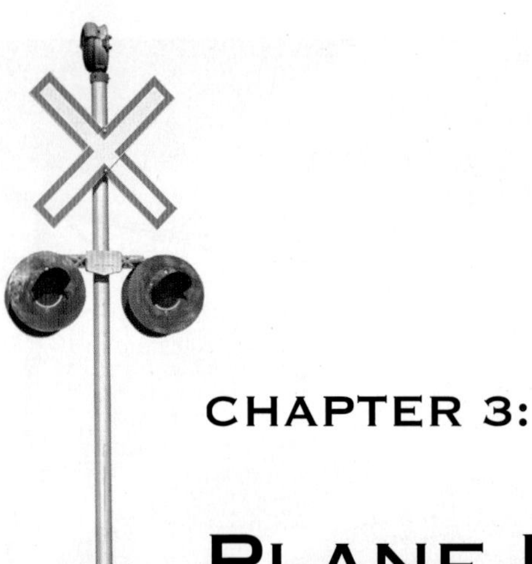

CHAPTER 3:

# Plane Figures: Anything But Plain

Floor tiles form shapes, or figures. Windows form shapes, too. A train engineer uses many dials. The dials are different shapes. Train conductors spend much of their day looking at figures. The tickets they collect are shaped like rectangles. Conductors often punch a round hole into the tickets. This circle tells the conductor when and where a passenger leaves the train.

Riders on a train can see plane figures inside the train. They see plane figures in the scenery outside, too.

**Quadrilaterals can be found on this train bridge in New York.**

Steel train bridges are an ideal place to spot plane figures. Steel bridges are very strong and sturdy. Over the years, steel bridges have replaced older, wooden ones.

Quadrilaterals are one type of plane figure. You can find them on steel bridges. Rectangles and squares are quadrilaterals. A rectangle has four sides and four right angles. A square is a rectangle with four sides the same length. It also has four right angles.

**This is Union Station in Washington, DC.**

Plane figures are everywhere at Union Station in Washington, DC. Many train stations across the country are called Union Station. A union station is a very large station with many train tracks and companies.

There are many examples of plane figures in this photo. Look carefully from floor to ceiling. Notice everything from the squares in the floor tiles to the octagons in the ceiling.

Look for geometric figures the next time you travel by train. What plane figures do you see in the station? Before you climb aboard, find acute, right and obtuse angles. Where do you see designs that look like parallel or perpendicular lines?

Check the train car for more plane figures, lines, and angles. Take a moment to look out your window. Can you see geometric figures in the scenery? Take a trip on a train or visit a station. You will see geometry in action!

**This mural is displayed at Union Station in Los Angeles, California.**

# Glossary

**acute angle** an angle that measures less than a right angle

**architecture** building designs and making structures

**intersecting lines** two or more lines that cross at exactly one point

**obtuse angle** an angle whose measure is greater than the measure of a right angle but less than a straight angle

**point of intersection** the exact point at which lines cross each other

**quadrilateral** a polygon with four angles and four sides

**rhombus** a parallelogram with four equal, or congruent, sides

**right angle** an angle that forms a square corner

Photo credits: cover, pp. 5, 6, 7 courtesy of Metro-North Railroad; pp. 1, 2, 8, 12 © Hemera Technologies Inc.; p. 3 Mario Tama/Getty Images; p. 4 © Richard Cummins/Corbis; p. 10 © Kelly-Mooney Photography/Corbis; p. 11 © Lee Snider/Photo Images/ Corbis; p. 13 © David Zimmerman/Corbis; p. 14 © Catherine Karnow/ Corbis; p. 15 © Ted Streshinsky/Corbis.